2018 Alachua County Hurricane Michael After Action Report / Improvement Plan

Gordon L. Gibby KX4Z

and the volunteers of the Alachua County ARES group

Volunteers Susan Halbert PhD, Leland Gallup JD, and Shannon Boal getting ready to drive to Panama City

Copyright © 2018   Gordon L. Gibby MD KX4Z

All rights reserved, **except the work may be copied with attribution to improve and enhance amateur radio emergency communications preparedness.**

ISBN: 9781729341919

Note: A Draft of this document, and the events recorded therein, was amicably discussed at great length by the participants and the Section Emergency Coordinator at a face-to-face meeting over lunch, in Gainesville Florida, on Saturday November 3rd, 2018. Participants all had access to the draft document. Comments and changes have in general been accepted from all participants, though the final editing was done by Gordon L. Gibby.   The document was then reviewed by the North Florida Amateur Radio Club on Nov. 14th 2018. Certain amendments were made at that time in view of discussions which had been held, and the amended document was heartily approved without opposition.

We are grateful for all the people who contributed to writing this
After Action Report / Improvement Plan.

DEDICATION

This After Action/Improvement Plan document is dedicated to all the volunteers at every level who worked so hard to help people impacted by Hurricane Michael, and to the cause of improving emergency response efforts by amateur radio in the future.

CONTENTS

Chapter	Title	Page
	Acknowledgments	vii
1	Introduction	1
2	Alachua County Merged Timeline	3
3	Observations, Recommendations, Alachua County Improvement Plan	15
4	APPENDIX: Alachua County Deployment Volunteers' ICS-214's	23
5	APPENDIX: KX4Z ICS-214	33
6	APPENDIX; Susan Halbert's Narrative	51
	About the North Central Florida Amateur Radio Club	57

(This page intentionally blank.)

ACKNOWLEDGMENTS

The North Central Florida Amateur Radio Club would like to acknowledge all the work done by the deployed volunteers and other officials who labored to respond in the wake of Hurricane Michael, and especially the reports submitted by Alachua County volunteers to help document the events and issues encountered.

Alachua County volunteers constructing antennas at a hurricane shelter.

(This Page Intentionally Blank)

1 INTRODUCTION

Hurricane Michael seemingly developed overnight and suddenly became one of the most massive hurricanes ever to impact the USA in history, reaching Category 4 and almost Category 5 status. Alachua County amateur radio volunteers were in the midst of final efforts just before their Simulated Emergency Test as this hurricane developed. Day by day, the forecast impacts became more dire, forcing a complete change of plans from a Simulated Emergency to a real one. The estimated total financial damage by the hurricane may reach $30 Billion, and insured losses are in the $6-10 Billion range.[1]

The Alachua County emergency-oriented amateur radio group has become stronger in recent years but still lacks significant exposure to state-wide emergency communications response efforts. Indeed, one of our volunteers indicated this is the first requested statewide deployment request in his memory. As such, the group was unfamiliar with interactions with Section and State officials, with whom we have very little experience in any simulation exercises. One of our mentors, Steve Waterman, commented that it is always less stressful to deploy in emergency situations with familiar people! That proved true in this instance as well.

The documentation in this review is intended to provide the most factual information on the events and a review of ways to improve response in future events.

---

[1] https://www.forbes.com/sites/adammillsap/2018/10/16/the-road-to-recovery-from-hurricane-michael/#32e74f762f6a

(This Page Intentionally Blank)

2   ALACHUA COUNTY MERGED TIMELINE

The amateur radio station established at the Bozeman Shelter.

Date / Time	Event
August / September 2018	North Florida Amateur Radio Club (NFARC) planned a Full Scale Exercise as part of the annual ARRL Simulated Emergency Test, to occur early in October. Because of a home college football game that would require the participation of local law enforcement / firefighters, the Alachua County exercise date was moved from the traditional first weekend to Saturday Oct. 13 to allow Region 3 MARC radio unit interoperability participation.

	Leland Gallup AA3YB and Susan Halbert KG4VWI were the volunteer planners for this exercise. They obtained permission for the use of the Gainesville Senior Center, the Alachua County EOC radio room, and a field at the Santa Fe College; provided required insurance documentation; created a master scenario, an ICS-201 incident briefing and ICS-205 frequency chart among other ICS documentation.
Sunday Oct 7	Newsweek reported that Tropical Depression 14 was expected to become Tropical Storm Michael later in the day, and was forecast to reach hurricane Cat 2 or 3 and cross the Florida Panhandle. Ref: https://www.newsweek.com/hurricane-michael-path-2018-landfall-timing-forecast-where-hurricane-headed-1157093
Monday Oct 8 1027 EDT	As Hurricane Michael came into existence and track estimates suggested it would hit the Florida Panhandle, the MARC unit leader regretfully bowed out of the planned Exercise due to expectations of government asset deployment.
	Now 5 days before the planned Exercise, Gordon Gibby sent news releases to local news media (a scoring item for ARRL SET participation), announcing the exercise. Copies also went to section, division and national ARRL leadership. No comments were received.
Tuesday Oct 9	Newsweek reported that Hurricane Michael was rapidly intensifying, expected to generate "dangerous storm surge, flooding and winds" -- making landfall as a Cat 3 hurricane. Storm surge might reach 12 feet, and landfall was expected during Wednesday in the Florida Panhandle. Alachua County was not within the prediction cone.
	Susan Halbert KG4VWI's state employer was shut down in preparation for the coming hurricane; from the Alachua County EOC, she worked to find possible volunteers to staff the Easton-Newberry shelter for possible hurricane refugees --- but that facility has concrete walls, which do not allow for antenna feedline placement. For over a year, plans to create a feedline pathway and add fixed antennas were not carried out. Progress was made in an effort to make officials aware of the need for antenna infrastructure in the Easton Center, especially since it often is the first to open in the event of an emergency.
Wednesday Oct. 10	More than 375,000 persons have been ordered to evacuate the Panhandle. The Hurricane made landfall as a Cat 4 hurricane (almost a Cat 5). https://weather.com/storms/hurricane/news/2018-10-09-hurricane-michael-preparations-impacts-florida-alabama-georgia Massive damage occurred in the Panama City area of the Florida Panhandle.
1900 (EDT)	With the Red Cross using their building for hurricane planning, NFARC met at the County Library headquarters at 7 PM. Leland and Susan advised that the hurricane was destroying all plans for the Exercise and that there was likely a significant need for actual ham radio response. The group agreed to postpone the Exercise to a later date, likely in January 2019, to be determined.
Thursday Oct 11 Morning	As daylight returned, the level of the devastation of Mexico Beach and other cities began to become evident. Gordon Gibby issued a corrective news release to media channels indicating that

	the Exercise had been postponed and that local amateurs were volunteering to the State, and noting that the ARRL encourages training and response to such emergencies.
1235 EDT	Karl Martin, the North Florida ARRL Section Emergency Coordinator, posted notice on the Florida EMCOMM Facebook page, of an official request from the State EOC for operators to volunteer; requiring 7 day commitment and vetted (background checked) by their local EOC. There was a significant discussion on that Facebook page amongst several commentators about the background check issue, since some counties don't routinely background check their volunteers. Some commentators said this had *always* been a requirement.
	Jeff Capehart, Alachua County EC had received an email from Karl which did not specifically require background checks by the local EOC. Alachua County ARES members are not currently routinely officially background checked by the EOC, although in the past there have been various background checks.

There were a lot of events, discussions, and news happening hour by hour as news of the devastation rolled in and local amateurs attempted to plan for a possible deployment. Initial deployment requests included the requirement for volunteers to provide not only their own food & water, but also **their own shelter.**

Efforts therefore began to prepare a privately owned travel trailer equipped for ham radio communications for possible loan to support volunteers if they were called to the Panhandle. Considerations included the electric brakes on the trailer requiring electric controls on the tow vehicle. However, later volunteer requests no longer indicated that they would need to provide their own shelter. |
| **Friday Oct 12**
1354 EDT | Email from Karl Martin indicating the deployment for 3 Alachua County volunteers is a "go";

'Deployment is a GO. The following is all the information. You can decide who should go to witch shelter. Please let me know ASAP who is going where. Please print multiple copies of there documents in case you need to give them to officers. ALWAYS keep the SERT Mission document with you. This is the mission tracking number and in a way your ID/Pass.

Shannon, No need for large gen.

Make sure to keep my information handy in case you have issues. I will listen to the SARNET, call me phone, text, email or on the SARNET anytime.

SARNET MAP
https://www.sarnetfl.com/uploads/6/1/7/0/61701057/180330_sarnet_map.pdf

Name of Shelter Address City Zip Shelter Manager Shelter Manager Phone

     Bay County

Leland Gallup AA3YB |

	Susan Halbert KG4VWI Shannon Boal K4GLM Deane Bozeman School 13410 FL-77,Panama City, Fl 32409 Sandy Sanders 850-348-2979 Northside Elementary School 2001 Northside Dr, Panama City, Fl 32405 Larry Messinger 314-330-8621 Rutherford High 1000 School Ave, Panama City, Fl 32401 Shella Florence 850-348-1667" Our three Alachua County volunteers are now assembling all their food, water, and fuel for a 7 day deployment, packing 3 pickup trucks with gear and radios.
Friday evening	Due to the late hour and the expectation that they would arrive in darkness (undesirable due to expected road debris and damage) the team decided to delay departure until 5 AM on Saturday
Saturday Oct 13 0500 EDT	Susan, Shannon, Leland and Gordon rendezvous at Exit 404, I-75,for last minute fuel can / fuel / digital equipment provisioning. Depart at 05:20
0730 EDT	Gordon Gibby sent out a final news release about the local volunteers heading to the Panhandle, and sent the release to ARRL officials as well.
0800 EDT	Susan, Shannon, Leland stopped in Tallahassee for gas and food; Waffle House: strictly limited menu; gas hard to find; Waffle House index yellow.
1230	Volunteers Arrived at Northside Elementary Shelter; ATT phones working intermittently;, no AC, no water or sewage. Shelter manager (Angie) suggested they go to Bozeman, where there was greater need (no communications). Took one vehicle to scope out Bozeman school.
1330	Volunteers Arrived at Bozeman shelter (K-12 school). 2 people from Lee Co EOC met us with animated request for communication support (Caitie Eck, John Kelly). They thought that we were the folks that they had ordered that morning. We explained that we were ordered via the Red Cross for shelter support. Red Cross shelter manager John Sanders 'Sandy' was concerned that these orders be deconflicted, because he was afraid that when the state saw that there were two similar orders, one would be canceled. We called Karl Martin, our ARES supervisor. He was made aware of the two similar orders, and that there was a need for both. He agreed that we should stay at Bozeman. We were also asked to check out the Bay Co EOC; Shannon had requested posting at the State EOC, so if the Bay Co. EOC needed help, he could go there.
1400	We proceeded with arrangements for operating facilities at Bozeman school. Kim Timmons, the assistant principal, helped us. We were not able to use the communications room because it was being used as a temporary morgue; however, there was a room nearby with ideal facilities. It has running water, power, and even a shower. "Mr. Jeff" (father was W4TAH), the school handyman, found us cots, bedding, and other supplies. We had received several versions for the address of the Bay County EOC. Kim knew where it was and gave good directions. It is not far from here (7 miles).
	Days earlier, volunteers from Gordon's church had headed to the Panhandle as part of the Baptist Disaster Relief. During Saturday afternoon, Gordon became aware from Marvin Corbin, of Baptist Disaster Relief of the need for private ham radio

	operators to fill a communications gap at Hiland Park Baptist Church in Panama City. Gordon and Nancy debated whether or not they could reach that area in time to be of any service, and still reach their jobs on Monday morning. Gordon emailed other NFARC members (who were not suitable for service through the ARRL channels) of the Baptist request.
1500	Left Bozeman to collect our vehicles, reconnect with Angie at Northside. She agreed with us that we should go to Bozeman. Traffic and debris made travel difficult all day. Roads reduced to one lane each way or less; side roads were impassable or filled with vegetation, downed wires, and other hazards; parking lots were infested with nails to puncture your tires (no mishaps for us!).
1830	Returned to Bozeman. Unloaded equipment. Too late (and dark) to put up antennas. [Note: see improvement plan for faster-assembly antennas, Observation 2.]
1930	Checked in with night shelter staff.
2015 EDT	Volunteers checked in with Gordon Gibby by telephone. (AT&T was working at Bozeman Shelter via AA3YB cell phone) Gordon Gibby emailed Karl Martin with their status update information. [see email later in appendices]
	Based on the 9 hour transit time reported by the team to reach the disaster area, Nancy and Gordon Gibby concluded it would be impossible for them to reach the Hiland Park Baptist Church with any usable time to benefit that group.
2059 EDT	Gordon Gibby sent an email to Karl Martin explaining the private request from Florida Baptist Disaster Relief for help with THEIR communications, explaining there were no further NFARC / ARES people in Alachua County who could meet the state requirements.
2115 EDT	Gordon Gibby sent an email out to local NFARC and friends relating the information received by telephone from the three volunteers, and copied Karl Martin. The email is in appendices and is reprinted here:

Passing along information received from our team tonight by telephone.

Incredibly difficult merely traversing small panama city; sirens everywhere. They think every policeman in Florida must be there! All three have settled for the night after 9 hours or more of travel, at Bozeman school northwest of Panama City after being told the need at another assigned shelter, which they reached, was not as great.

They have not yet got any antennas up. Pine trees snapped off in the middle; no tops. (Kinda what Steve Waterman always told me) they are eyeing a tower they think they can exploit.

Bozeman shelter being used for special needs (extremely ill). Of note, Allan West is deployed somewhere out there also with the FEMORS group. Not sure where (forgot). |

	AT&T works right where they are, but extremely spotty. In town – much less connectivity. Biggest problem they are addressing is that the county EOC seems to have no connectivity with Bozeman shelter so they are planning to tackle that tomorrow. I read elsewhere that Verizon is working very very hard & steadily. The team thinks that once phones work there will be much less need for them. *They were lucky to have cell phones with service. Future note; take cell phones from multiple vendors so if you're lucky, one will work!!* *They are unsure whether they can make it back to assigned shelters inside Panama City. Traffic, road conditions, etc. Doing the best that they can. Incredibly difficult to even move around modest distances.* *But all safe, with real electricity and cots and comfort for tonight.* *I don't have a phone number for Debbie Boals but I've checked with Angela Gallup and she has gotten word from her husband.* *I wouldn't go sharing any of this with media......* *Based on their 9 -hour trek to even get there....there is nothing that I can do with the one day I have remaining available.* *Karl Martin has been doing great things and the team has really appreciated working with him. I'm recruiting him to speak at our February Conference.* *Cheers,* *Gordon Gibby* [italics added]
2203 (EDT)	Karl Martin sent an email to Gordon Gibby indicating that the last contact Karl had had with the team they had reached a shelter and were planning to get Shannon to the EOC as an EOC station and then Susan and Leland planned to go back to shelter – Karl asked me to send him an email message to explain how to reach the team by WINLINK.
2252 (EDT)	Gordon Gibby found the email timed 2203 from Karl Martin and replied, explaining how to reach WINLINK addresses using ordinary email. (copy of email is included in later appendices) Gordon then went to sleep
Sunday, Oct 14 0730 CDT	Alachua County volunteers set up stations. Push-up pole on a tripod with UHF, VHF station at the top, US flag, Buckmaster antenna on a pulley
0902 EDT	Karl Martin replied to Gordon Gibby thanking him for the information.

	Gordon and Nancy were off to church and later on the way to see their son and grandson in Orlando.
1008 CDT	Alachua County volunteers established UHF phone communications with SARNET through the Chipley station at 10:08 CDT. We were their first Bay County contact.
1030 CDT	Alachua County volunteers established HF phone communications with NFAN on 3950 kHz.
1100 CDT	Alachua County volunteers established HF digital Winlink communications on 40 meters. Sent one message; received 7.
1125 CDT	Alachua County volunteers replied to one of the health and welfare messages. Received request to check on Hiland Baptist Church, from Red Cross, to assess their communications needs.
1130 CDT	Alachua County volunteers touched base with the staff at Bozeman the Health Department staff (special needs shelter) said it was a horrible night. Staff we spoke with at shelter said that they had "no communications" with the EOC. This explained why we had been tasked to provide communications with the EOC from the shelters (shelter counts, needs, etc.). AT&T phones were working, but possibly there often was not a response from the EOC, and routine messages had to be sent by courier, which could be a 2 hour round trip. It seemed to us that in order to accomplish our task, it was essential to find out how to obtain communications with the local Bay County EOC.
1135 CDT	Alachua County volunteers received message from Gordon that WINMOR was not working [GLG: at KX4Z alone], so test message sent on Winlink VARA
1135 CDT	Alachua County volunteers established VHF phone connection with local repeater (later learned on 800 ft. tower) 145.330, tone 100. We listened quite a bit on this repeater and learned that there was no contact with the Bay Co. EOC.
1230 CDT	Continuing to make contacts on our station. Shannon went to see the local Bozeman Red Cross staff to advise them that we were going to the Hiland Baptist Church. Mr. Sanders was not available, but Shannon spoke with James "Doc," a Red Cross worker. Doc went to a "no admittance" office while Shannon waited a long time but never obtained access. (He learned that the Red Cross was making the Deane Bozeman school their primary location, hence the private office.) The Red Cross was moving Special Needs patients to Tallahassee to make room for their headquarters at the school. People were arriving from Washington DC. We again were asked to go to Hiland Baptist Church to assess communication needs.
1230 CDT	Alachua County volunteers traveled toward the Bay County EOC to figure out how to obtain communications there. This took considerable time on road; we made decision to have all three go with the possibility of leaving one of us there at the EOC to provide the communications link with Bozeman. (Karl Martin had suggested posting Shannon there.) We could not find a person at the EOC able to identify who we should talk to about radio communications at the EOC; the "ham" we were informed was there in fact was not there; we were told to return at 19:00 when Bob Edmunds, the EOC "operator," would return for the night shift. LEO on duty said that their radios mostly didn't work. AT&T cell phones were the best option.

1430 CDT	Went to Hiland Baptist Church; arrived about 16:00. This took one and a half hours on the road. The Baptist Church has a dedicated high end communications trailer, equipped with a HAM radio that will do all bands (ICOM 706 Mark II G). Tim (KD5SSF) is an experienced operator with the Southern Baptist Disaster Relief, especially on HF, but there was no HF antenna. We helped him set up the radio to communicate with the local 145.330 repeater. We also told them they could probably reach us directly at Bozeman. The Hiland operation was a food distribution operation, not a shelter. Having satisfied the requirement we were given, we had supper and left.
Afternoon	Nancy and Gordon Gibby visited their grandchild and son in Orlando on Sunday and on the trip back tried to listen to UHF repeaters of the SARNET. At approximately 3:30-4PM [EDT] we heard Leland AA3YB many many times calling on the SARNET trying to establish communications. The net control heard him many times and responded many times – it appeared that Leland could not receive the replies that were being sent to him. Nancy and I were concerned that there was a communications issue, distance problem, antenna problem, something.
1655 EDT	Gordon Gibby forwarded a message to Karl Martin that he had just found a Winlink email from the Alachua County team (they had sent it out at 20:36 GMT) The message is appended later in an appendix In that email, Leland explained they had reached the Bay County EOC but found no radio operator there; that they were to return later in hopes of reaching the radio operator, and that they would keep Karl informed.
1705 EDT	Leland send Gordon an email via Gmail (via AT&T cell) explaining they were going to try to reach Hiland Baptist at the request of the Red Cross. That email is appended.
1815 CDT	Alachua County Volunteers Left Baptist Church to go to the EOC. It took 2 hours to drive there due to bad traffic. Roads were hazardous due to downed power lines and poles, pieces of demolished buildings, broken glass, large amounts of vegetation from broken trees, convoys of emergency vehicles, traffic accidents, non-functional traffic signals, and general traffic congestion.
Evening, EDT	After Gordon Gibby returned home from Orlando, he tried to monitor both SARNET and 3.950 on two radios in my living room. Trying to listen to both SARNET and HF 3.950 simultaneously. At some point Gordon heard Susan's voice on one or the other. Gordon gave up listening to the SARNET and tried to check in unsuccessfully to the 3.950 net --- so went upstairs to a better antenna/transmitter and checked in. There were discussions of why there were so few volunteers, the opinions expressed were it had to do with ICS requirements. Gordon notified the Net Control Station N0TW that anyone could reach the Alachua County team using WINLINK. N0TW indicated that was not useful as he didn't have WINLINK. Gordon was caught off guard and didn't know how to reply. Several minutes later Gordon composed an explanation of how anyone using any email can email anyone using WINLINK by addressing CALLSIGN@WINLINK.ORG and putting //WL2K as the first item in the Subject line – and N0TW was very grateful for the explanation and quickly sent me an email to my WINLINK account to test this. Gordon received his email and replied, giving all the Alachua County Winlink email addresses. There were others on the net who also seemed surprised that this could

	be done. Note: this was the same information explained to SEC Karl Martin the previous evening. As the misconceptions of the difficulties with volunteers continued on the net, Gordon was able to explain that Karl Martin had put in writing all the exact requirements and that they were the First link on the section web page – but then he was told that people didn't know where the section web page was....Gordon was again caught off guard and suggested they Google North Florida ARRL Section – at that point Kevin Bess came on frequency and gave the entire URL. At 8:13 PM EDT Gordon sent an email to Karl Martin documenting those events. The email is appended.
2020 EDT	Gordon received a phone call from Karl Martin asking about what Gordon knew about the Alachua County team's efforts and apparently in particular why they had visited the Hiland Baptist Church. They spoke for (estimated) 20 minutes. Gordon can't remember exactly what he told Karl but immediately after the phone call, looked it up and minutes later that evening Gordon sent Karl an email that Gordon had received from Leland explaining they had been requested to do that by the American Red Cross
2015 CDT	Alachua County volunteers Arrived at Bay County EOC. (for the second time) Met Charlie Wooten (NF4A) by coincidence, as he was leaving.. He owns the 145.330 repeater and runs several local radio stations. He has been in the local HAM community for many years. He is somewhat discouraged by the lack of interest in emergency communications in the local HAM community. He was not at the EOC as an amateur operator, but instead in his professional capacity with responsibility to have local radio stations up and running. (Learned tonight on NPR – 22 October – that the local PBS station has set itself up in the EOC to provide public information.) Met Bob Edmunds, the HAM who had checked into the 3950 net from Bay County. He works for the EOC, but he is assigned to GIS and traffic, not radio. There was no known amateur radio operator or any ability to operate at the Bay County EOC at that time. We signed in as a resource with Gerald Pangelinan, the Resource Unit Leader. We learned that the radio room was being used by the Air National Guard. The HF antenna has been repaired by the National Guard, but is in use for their mission. There is no place to put any antenna, and no place to operate a HAM radio at the EOC. The Radio Systems Administration team called the State EOC and found out that our task had nothing to do with the Bay County EOC, but only the State EOC in Tallahassee. We attempted to find actual need by briefly talking with the IC. but he was swamped with many inputs from a host of actors reporting on scene from government, law enforcement, fire and rescue, communications, industry, and private organizations. We repeatedly heard that there was no known request from the state EOC for amateur radio support at the Bay County EOC; only the Red Cross had requested support. Moreover, all the potential amateur radio facilities at the EOC had been re-purposed for other more urgent needs (National Guard). Consequently, no apparent request, need, or infrastructure for amateur support existed at the EOC.

| | Met Gary Huckabay, recently arrived Red Cross liaison supervisor for the night shift. After discussion, it became clear that any remaining communication problems with the shelters could be solved by buying some AT&T "burner" phones. After further discussion with Gerald and Gary, it was determined that we would wait for a day to determine if our services were needed elsewhere (staging area for the County, for example), but with respect to the Red Cross, the need for our services was resolved: the Red Cross understood that there was a simple method for communicating with the shelters (AT&T phones), and the EOC had no apparent need for or desire for amateur radio support. We made it quite clear that we were capable of being re-tasked to support other locations as needed, but this was a time consuming effort for us given the difficulties of road travel. Rutherford Red Cross shelter is being closed tomorrow (problems there), and Bozeman shelter and the Northside shelter can communicate by AT&T.

We asked Gary Huckabay about what to do about the requests for information about missing people in Bay County. We were told that they could look on line in the Red Cross "safe & well" site. We mentioned that these people did not have any access to the internet. It was then suggested that the relatives call the Bay County sheriff.

In spite of the perception of a lack of need for communications between the shelters and the EOC, we did notice some discrepancies between the information that the EOC was receiving and reality; for example, they were not aware that anyone had died at the Bozeman shelter. They also thought that the counts they received were exact numbers rather than estimates. These problems should resolve as the AT&T network becomes more robust. |
|---|---|
| 2200 CDT | Departed EOC and returned to Bozeman operation. |
| 2230 CDT | Arrived Bozeman shelter very tired. Leland sent email to KX4Z apprising of the events and with Susan recorded day's events on 214. |
| 2400 CDT | Alachua County volunteers lights out. |
| | |
| **Monday, Oct 15** 0237 EDT | Gordon Gibby awakened in the night and found an email from Leland written at 0053 (EDT) and forwarded it to Karl Martin at 0237 EDT The email is attached as an appendix. The Alachua County volunteers described their efforts of the day
-- first visit to Bay County EOC – no radio operator, told he would be back at 7 Pm
-- second visit to Bay County EOC – 8 pm, still no radio operator, but they figured out that they could communicate using AT&T phones. The email lists their points of contact.

Team found out that all the requested shelter/posts in their area could communicate by simply getting AT&T telephones. |
| 0645 CDT | Leland Gallup spoke with Gerald Pangelinan (ATT cell phone), trying to find if there had been any development in re-purposing/requesting their re-deployment to a different location that there was no further tasking request for the Alachua County volunteers from the State EOC or others locally, indicating no further need. |

0700 CDT	The Alachua County volunteer team spoke with Karl Martin about the events of the previous day. There was a disconnect between what we had observed in Panama City and what people outside of the area perceived. **Mr. Martin was under the impression that there was an operator at the Bay County EOC**, and that the "Director of the EOC" wanted amateur radio communications. We attempted to explain that AT&T phones were providing increasingly adequate communication, and that **there was no amateur radio operator at the Bay County EOC to receive any shelter communications via amateur radio (nor was any such position likely to be staffed, because both the office and the equipment were in use by the National Guard!)**. Moreover, the IC did not need or want HAM radio communications. In reality, our assigned mission was no longer needed due to re-establishment of cell coverage; however, the conversation with Mr. Martin made an unpleasant end to an otherwise successful adventure.
0800 CDT	Took down antennas and packed. This took about 4 hours.
1200 CDT	Alachua County volunteers met with local shelter Red Cross shelter managers and received warm thank-you for showing up (never did meet the chiefs from National Headquarters).
1215 CDT	Shannon fell on his back attempting to move a poorly secured pallet of waste cardboard placed in the road by some of the National Red Cross workers. He was injured, but not severely. "Mr. Jeff," the chief custodian from the school, provided a bandage, disinfectant, water, aspirin, and Gatorade. GLG: *The volunteer group noted there were some staff changes at the shelter occurring but in the interests of privacy as suggested by K.M. I've deleted those.*
1230 CDT	Filled Susan's and Leland's cars with gas from our gas cans; left for Gainesville.
1500 ?EDT	Stopped for lunch in Tallahassee (first meal of the day) at Steak & Shake. Limited menu, but AC and lights on. Gas is still scarce. Filled Susan's car with gas from our gas cans.
1600 EDT	Found a station with diesel fuel for Shannon. They had no regular gas
1730 EDT	Alachua County volunteers Arrived to the outskirts of Gainesville.
1925 EDT	Gordon Gibby received an email from Karl Martin wanting to be sure the team arrived home safely. Gordon replied they had, and invited Karl to call at his convenience to discuss these events in greater detail so that Gordon would have heard all sides.
Tuesday, Oct 16 morning	Shannon Boal had a discussion with Karl Martin about the previous events. Shannon invited Karl to the evening Gainesville Amateur Radio Club meeting.
Tuesday evening Gainesville Amateur Radio Society regular	Leland conducted an after-action "hot wash" of the deployment. This was followed by an animated discussion among the members present at the regularly-scheduled Gainesville Amateur Radio Society monthly meeting.

meeting 6:30 PM	
Days Later	Gordon Gibby reached and interviewed the EC of Bay County at his evacuation location on the east coast of Florida. The volunteer had limited time to organize ARES operations in Bay County because of work responsibilities, and had experienced very low interest from club members in the County ("when ARES came up, people would stand up and leave!") and had also had little success in building interoperability relationships with the Bay County EOC.
Sunday evening, Oct 28 2018	Gordon Gibby learned in a telephone conversation with Karl Martin two key additional pieces of information: 1) multiple additional ham radio teams were sent to the Bay County EOC and were turned away 2) Karl tried several times on Sunday Oct 13 to reach the Alachua County team, utilizing voice telephone and text telephone and had circuit failures and similar difficulties and was unable to establish communications with them using telephone.

3 OBSERVATIONS, RECOMMENDATIONS, & ALACHUA COUNTY IMPROVEMENT PLANS

Trees left by Hurricane Michael

Note: The Alachua County NFARC group can only make improvement plans for its own local efforts. Therefore "Recommendations" are presented for not only Alachua County but for much broader usage, while the Improvement Plan is directed solely at Alachua County efforts.

Observation	Recommendations for amateur radio operators	Improvement Plan (for NFARC / Alachua County ARES)
1. Lack of EOC-background checks hindered many potential volunteers. Discrepancies and changes in volunteer requirements impaired additional potential volunteer response. • Initial published volunteer requirements included self provision of shelter. What form of background check and who must have completed it was different in different communications. • The North Florida Section Emergency Comms plan indicates FDEM will require only a driver's license and will perform any additional checks required. It does not indicate that prior EOC-completed background checks are required. [2] (Ref: http://arrl-nfl.org/wp-content/uploads/2016/03/NFL-Section-Comm-Plan-FINAL-May-1-2016.pdf)	• Recommend that each county volunteer group arrange to acquire and maintain background checks by their local EOC sufficient for state acceptance. • Recommend that each county volunteer group formally induct all members into their RACES plans. • Recommend that the North Florida Section Communications plan be updated to specifically encourage EOC-background checks as were required in this instance.	• Sign all potential volunteers into the RACES plan. • Request local EOC to background check all potential volunteers and to repeat it every two years. • Standardize out-of-county deployment roster and out-county deployment gear lists. • Standardize set of pre-printed handouts with plain-language text detailing what services amateur radio operators would provide, where, and how. • Ideally, the deployment should function in an ICS format, so that lines of authority and mission expectations are understood by all parties. Clearly, things might need to change once the situation on the ground is assessed, but if lines of authority are clear, new information can be processed and addressed within the ICS plans process.
2. Delays in deployment of volunteer amateur radio operators reduced their effectiveness to the distressed area during the window of greatest need. Final deployment approval and commissioning did not occur until Friday afternoon for Alachua County volunteers, 2 days after the arrival of Hurricane.[3] Due to the long and uncertain travel, volunteers then departed at 0500 the next morning. By comparison, local church	• Recommend that leadership at multiple levels pre-vet possible volunteers, stage equipment and supplies before arrival of the hurricane and be ready to deploy immediately when travel becomes safe. • Consider staging possible pre-vetted volunteers to move closer to the distressed area if there are delays in formal approval. • Consider offering assistance to private / NGO groups with legitimate needs and more	• Investigate possibilities of service to NGOs such as the SATERN and Florida Baptist Disaster Relief organizations. • Ask potential volunteers to pre-arrange radio, food, water, and other supplies so as to be more readily available for service. • Include equipment that can be rapidly assembled and broken down, so that operations can be moved rapidly as needed. Particularly the ANTENNAS. (STRONG emphasis by Leland

---

[2] From the NFL Section 2016 Emergency Communications Plan, page 5"" To qualify for official deployment requires you to provide your Florida Driver's License ID number or other acceptable government issued photo ID. Deployment for emergencies under mission requests from the FDEM will not be authorized unless the amateur radio operator has voluntarily submitted the necessary information and has been cleared by that agency"

[3] Operations changed the plans to become County-based rather than alphabetic Corridors in order to provide better service, and this resulted in some delay to amateur radio deployment. personal communication, Phil Royce

volunteers supporting the Florida Baptist Disaster Relief had mobilized on Thursday, 1 day after the hurricane. By the fourth day after the hurricane (Sunday) our volunteers documented that communications were possible throughout their requested service locations simply by using AT&T cell phones. [GLG: Caveat: cell phone switching for incoming phone calls was not yet working perfectly based on Karl's experience.]	nimble response as they may have fewer resources than the professionals in state government.	Gallup and Susan.)
3. Local Alachua County press releases were not reviewed by more experienced Section leaders (Press release resulted in a published story before a comment from a Section official was received.)	• Send all press releases related to Section-wide events to more experienced Section leaders with a request for review in timely fashion.	• Send all applicable press releases to more experienced Section leaders with a request for review in timely fashion
4. Alachua County deployment volunteers did not remain in continual communication with emergency nets, but instead were in discontinuous communication. (However, by Sunday, they could potentially be reached everywhere by cell phone because at least one member of the team had AT&T cell phone and the team documented that this was working at all relevant locations.) [GLG: switching for incoming calls was apparently not yet working well.] Communications were sent out or attempted: 1. Saturday 1330 – telephone to Karl Martin 2. Saturday 2015 – telephone to G. Gibby, who forwarded to Karl Martin 3. Sunday 1008 – UHF (sarnet) 4. Sunday 1030 - HF (3.950) 5. Sunday 1100 - WINLINK 6. Sunday? 4:30 – failed UHF	• Whenever possible assign a volunteer to remain in VHF or HF contact (or both) so that continual communications are assured. • Better familiarization with deployment radios to avoid communications failures. • Be aware that cell signals don't prove working cell phone switching; send situation reports and test results to be disseminated through ICS-201 briefings so everyone is more aware of the actual state of cell communications.	1. During any deployment, attempt to maintain CONTINUOUS availability by radio. 2. Document in a contemporaneous ICS-214 all communications availability and any periods of non-availability and the explanation. 3. **Send these records by radio or WINLINK or EMAIL to the supervisor at FEMA intervals of 12-hour periods.** (There was some confusion about who should receive our ICS214 forms as we produced them.) 4. **Be aware that cell signals don't prove working cell phone switching; send situation reports and test results in the ICS-214 to be disseminated through ICS-201 briefings so everyone is more aware of the actual state of cell communications.**

attempts 7. Sunday 2036 GMT – WINLINK, which Gordon forwarded to Karl Martin 8. Sunday Midnight – email to Gordon, which Gordon forwarded to Karl Martin at 0237 AM 9. Monday AM 3950 (Susan) Despite these many communications, there were significant gaps of RADIO communications, and outside personnel were likely unaware that cell phone/text could potentially reach them -- GLG: Note the caveat – Although the Alachua County Team experienced strong AT&T cell phone signals, Karl was repeatedly unable to reach them on INCOMING calls to the area, getting busy signals and unusual messages. **Apparently the emergency reconstruction of the AT&T cell phone system worked better for outgoing phone calls than for incoming phone calls** – meaning amateur radio was still important.		
5. Alachua County volunteers were at times unaware of whose orders to follow, whose to ignore, and exactly what was their precise mission. • They never received an ICS-201 throughout >48 hours of deployment. • They never received an ICS-205 • They never received an ICS-204 With an initial (written, email) assignment to provide communications for Red Cross shelters, they responded to Red	• Radio amateur leadership can provide more effective leadership with ICS tools such as ICS-201, 204, 205 at 12 hour intervals as explained in suggested ICS courses 100, 200, 700, 800, 300, 400. Organization structure should be clearly delineated, with a clear chain of supervision and a clearly defined method for unifying effort on-scene and in the rear. • Assignments should be provided in written format and with as much detail and clarity as possible given the circumstances. • Supervisors should be immediately available (or	• Alachua County volunteers should request written ICS documentation for each period of service. • Maintain copies of all directives received • Keep contemporaneous ICS-214 • Notify supervisors of any confusion or possible change in circumstance or mission effort. • Clarify all ambiguities and seek clear guidance from a unified chain of deployment supervision.

Cross requests for communications with EOC and Hiland Park facility. Directive communications to the Alachua County deployment team were issued 1. Friday 1354 by email when they were in Alachua County. 2. Saturday 1330 in a telephone call initiated by the volunteers 3. Monday 0700, telephone call sending them home. 4. Additional attempts by Karl Martin to reach the group using voice/text telephone on Sunday Oct 14 were unsuccessful. (ref: personal communication, Karl Martin, Oct. 28)	replacements provided) for consultation. Ideally, new information is incorporated each operational period through the ICS Plans process. • Volunteers should be provided with standardized information concerning protection against radio operations personal liability.	
6. Emergency net participants and supervising officials were not sufficiently familiar with WINLINK systems to make useful communications with volunteers, although WINLINK capability was one of the factors in choosing volunteers. • Both the SEC and an HF emergency net control station gladly received information on how to reach deployed stations via WINLINK using ordinary GMAIL or other email. • Email is widely used throughout the world because of its ability to store messages for people who are "on the go" • The NFL Section Communications Plan heartily endorses WININK and encourages every county to have an RMS or	• All Section leadership and emergency communications should gain basic familiarity with how to access WINLINK emails, as this is a featured communications system advocated by the ARRL. • Section and County level emergency leadership should gain and maintain actual WINLINK email transmission and reception experience to become competent to handle traffic during the stress of an emergency. • Because of the destruction of local communications systems and the need for HF communications in this emergency, Section and Local emergency leaders should develop HF WINLINK competency.	• Alachua County NFARC will continue to offer educational opportunities such as the Emergency Symposium held in February 2018, and tentatively to be held again in February 2019

• digipeater method to reach one. • The WINLINK system handles approximately 50,000 messages every month by radio.		
7. Rumor control is imperative. Shannon Boal was made aware of a rumor that the team had argued with the Bay County EOC, which is simply not true according to Shannon. [GLG: In fact, additional amateur radio teams sent to that EOC were also turned away. The exact nature of the rumor in our case is not known and no written statement about it has been received.]	• Always fact-check rumors.	• Request that all rumors be verified.
8. Bay County ARES development was small and ineffective, and there were essentially zero relationship with the Bay County EOC The Alachua County team was not apprised of these facts and therefore did not recognize the issues presented by the Red Cross Shelter request to establish communications with the Bay County EOC. Bay County EOC radio facilities were utilized by the National Guard, with whom no interoperability communication frequencies were known to the Alachua County volunteers, and since the radio infrastructure had been reassigned to the National Guard, it essentially negated the opportunity to do classic shelter-EOC radio communications.	• Develop knowledge within the Section of which ARES groups have which capabilities. • Perform "gap analysis" to determine where weaknesses need to be strengthened • Make deployed volunteers aware of local radio interoperability issues prior to deployment through discussions with local amateurs / EC / AEC from those areas.	• Alachua County volunteers should always attempt to reach the EC and other officials of ARES(R) to which they are deployed to get valuable "local knowledge" to avoid interpersonal difficulties. • Alachua County volunteers should foster and maintain good relations and interoperability with Alachua County EM, LEO, and first responders so avoiding Bay County's lacking amateur operator support
9. With massive hurricane damage to trees and structures, Susan's advice to bring antennas that are	• Utilize telescoping masts and other self-supporting supports for antennas. As these may not be	• Encourage local members to increase their asset base of telescoping antenna masts, and

self-supporting appears wise. At one installation a very long coax run was deemed necessary to go from operating position to safe location for antennas.	very stiff, lightweight antennas may be desirable. • Bring plenty of coaxial cable.	coaxial cable. • Encourage antennas that are very quick to deploy and take down.
10. Flagging ("caution") tape was needed in some instances to protect people / antennas. Personal liability of volunteers for injuries caused by operations is a concern.	• Acquire and transport suitable caution tape.	• Acquire caution tape • Review the protections provided by FSS 768.1355 to determine if they are adequate for our volunteers
11. At times, the deployed group felt a printer would have been helpful.	• Find low-power (bubble-jet) printers that can be utilized in emergency situations.	• Acquire suitable bubble jet printers for use by deployment teams.
12. There appear to have been very little usage of the ARRL Radiogram during this emergency response. Tactical communications, email, text and telephone were more prominent.	• Continue to train volunteers to seek out all possible communications techniques. • Provide ICS-205A forms with contact information for all relevant deployed and support personnel.	
13. There appeared to be a significant number of emergency communicators who are unfamiliar with the Section Web page, and also were not aware of the published requests and qualifications sent out by the SEC. (The SEC emailed all ECs)	• Recommend that the Section seek to make members more aware of the Section web page. • Recommend that ECs develop improved methods to pass along written information from Section leadership to their volunteers.	
14. The amount of "rear support' required to appropriately support deployed volunteers is LARGE and must be considered in any deployment. Supervisory personnel (the SEC) appeared to be inundated with communications and exceeding recommended span of control, optimal being 5 (ref: https://emilms.fema.gov/IS100c/groups/29.html This may have contributed to sparse communications and directives.	• Recommend that adequate systems for field assignment of additional deputy or assistant personnel be created so that assistants can be appropriately empowered.	• Train volunteers to maintain a High index of awareness for instances where span of control is being exceeded, and move quickly to request adequate rear support and supervisory personnel so that the situational awareness of supervisors is not impaired.

.22

4 ALACHUA COUNTY DEPLOYMENT VOLUNTEERS ICS-214'S

Alachua County volunteers with a shelter manager.

UNIT LOG	1. Incident Name Hurricane Michael Panama City	2. Date Prepared 14 October 2018	3. Time Prepared 14:19
4. Unit Name/Designators Bozeman school shelter radio	5. Unit Leader (Name and Position) Shannon Boal		6. Operational Period **13 October 2018**

7. Roster of Assigned Personnel

Name	ICS Position	Home Base
Leland Gallup	radio operator	Alachua Co
Shannon Boal	radio operator	Alachua Co
Susan Halbert	radio operator	Alachua Co

8. Activity Log

Time	Major Events
05:00 EDT	rendezvous at Exit 404; obtained gasoline, other supplies from Dr. Gibby; made plans
08:00	stopped in Tallahassee for gas and food; Waffle House: good waffles, nice waitress; gas hard to find; Waffle House limited menu; Waffle House index yellow.
12:30 CDT	Arrived at Northside Elementary Shelter; ATT phones working intermittently; plumbing problems –, no AC, no water or sewage. Shelter manager (Angie) suggested we go to Bozeman, where there was greater need. Took one vehicle to scope out Bozeman school.

13:30 CDT	Arrived at Bozeman shelter (K-12 school).2 people from Lee Co EOC met us with animated request for communication support (Caitie Eck, John Kelly). They thought that we were the folks that they had ordered that morning. We explained that we were ordered via the Red Cross for shelter support. John Sanders 'Sandy' was concerned that these orders be deconflicted, because he was afraid that when the state saw that there were two similar orders, one would be canceled. We called Karl Martin, our ARES supervisor. He was made aware of the two similar orders, and that there was a need for both. He agreed that we should stay at Bozeman. We were also asked to check out the Bay Co EOC; Shannon had requested posting at the State EOC, so if the Bay Co. EOC needed help, he could go there.
14:00 CDT	We proceeded with arrangements for operating facilities at Bozeman school. Kim Timmons, the assistant principal, helped us. We were not able to use the communications room because it was being used as a temporary morgue; however, there was a room nearby with ideal facilities. It has running water, power, and even a shower. "Mr. Jeff" (father was W4TAH), the school handyman, found us cots, bedding, and other supplies. We had received several versions for the address of the Bay County EOC. Kim knew where it was and gave good directions. It is not far from here (7 miles). Angie was good with us leaving to work at Bozeman. Traffic and debris made travel difficult all day. Roads reduced to one lane each way or less; side roads impassable or filled with vegetation, downed wires, and other hazards; parking lots infested with nails to puncture your tires (no mishaps so far).
15:00	Left Bozeman to collect our vehicles, reconnect with Angie at Northside. She agreed with us that we should go to Bozeman.
18:30	Returned to Bozeman. Unloaded equipment. Too late (and dark) to put up antennas.
19:30	Checked in with night shelter staff.
20:00	Checked in with Gordon Gibby.

October 14

UNIT LOG	1. Incident Name Hurricane Michael Panama City	2. Date Prepared **14 October 2018**	3. Time Prepared 22:22
4. Unit Name/Designators Bozeman school shelter radio	5. Unit Leader (Name and Position) Shannon Boal		6. Operational Period **14 October 2018**

7. Roster of Assigned Personnel

Name	ICS Position	Home Base
Leland Gallup	radio operator	Alachua Co
Shannon Boal	radio operator	Alachua Co
Susan Halbert	radio operator	Alachua Co

8. Activity Log

Time	Major Events
7:30	Set up stations. Push-up pole on a tripod with UHF, VHF station at the top, US flag, Buckmaster antenna on a pulley.
10:08	Established UHF phone communications with SAR-net through the Chipley station at 10:08 CDT. We were their first Bay County contact.
10:23	Established HF phone communications with NFAN on 3950 kHz.
11:00	Established HF digital Winlink communications on 40 meters. Sent one message; received 7
11:25	Replied to one of the health and welfare messages. Received request to check on Hiland Baptist Church, from Red Cross, to assess their communications needs.

11:35	Received message that WINMOR was not working, so test message sent on VARA
11:35	Established VHF phone connection with local repeater (later learned on 800 ft. tower) 145.330, tone 100. We listened quite a bit on this repeater and learned that there was no contact with the Bay County EOC. They said that Bob (WB4BLX) had checked in on 3950 (but see below – his check-in had been from home, not from the EOC).
11:30	Touched base with the staff at Bozeman and; the Health Department staff said it was a horrible night. Staff we spoke with at shelter said that they had "no communications" with the EOC. This explained why we had been tasked to provide communications with the EOC from the shelters (shelter counts, needs, etc.). ATT phones were working, but possibly there often was not a response from the EOC, and routine messages had to be sent by courier, which could be a 2 hour round trip. It seemed to us that in order to accomplish our task, it was essential to find out how to obtain communications with the local Bay County EOC.
12:30	Went to EOC to figure out how to obtain communications there. This took considerable time on road; we made decision to have all three go with the possibility of leaving one of us there at the EOC to provide the communications link with Bozeman. We could not find a person at the EOC able to identify who we should talk to about radio communications at the EOC; that the "ham" we were informed was there in fact was not there; we were told to return at 19:00 when Bob Edmunds, the EOC operator, would return for the night shift. LEO on duty said that their radios mostly didn't work. ATT cell phones were the best option.
12:30-14:30	Continuing to make contacts on our station. Shannon went to see the local Bozeman Red Cross staff to advise them that we were going to the Hiland Baptist Church. Mr. Sanders was not available, but he spoke with James "Doc," a Red Cross worker. He went to a "no admittance" office while Shannon waited a long time but never obtained access. He learned that the Red Cross was making the Deane Bozeman school their primary location. The Red Cross was moving Special Needs patients to Tallahassee. People were arriving from national headquarters. Tasked again to go to Hiland Baptist Church to assess communication needs.

14:30	Went to Hiland Baptist Church; arrived about 16:00. This took one and a half hours on the road. The Baptist Church has a dedicated high end communications trailer, equipped with a HAM radio that will do all bands (ICOM 706 Mark II G). Tim (KD5SSF) is an experienced operator with the Southern Baptist Disaster Relief, especially on HF, but there was no HF antenna. We helped him set up the radio to communicate with the local 145.330 repeater. We also told them they could probably reach us directly. The Hiland operation was a food distribution operation, not a shelter. It was clear that they at least had one VHF means of communication after our station. Having satisfied the requirement we were given, we had supper and left.
18:15	Left Baptist Church to go to the EOC. It took 2 hours to drive there due to bad traffic. Roads were hazardous due to downed power lines and poles, pieces of demolished buildings, broken glass, large amounts of vegetation from broken trees, convoys of emergency vehicles, traffic accidents, non-functional traffic signals, and general traffic congestion.

20:15	Arrived at EOC. Met Charlie Wooten (NF4A) by coincidence, as he was leaving.. He owns the 145.330 repeater and runs several local radio stations. He has been in the local HAM community for many years. He is somewhat discouraged by the lack of interest in the local HAM community. He was not at the EOC as an amateur operator, but instead in his professional capacity with responsibility to have local radio stations up and running. Met Bob Edmunds. He is assigned to GIS and traffic, not radio. His check-in on 3950 had been from home, not from the EOC. There is no known amateur radio operator or ability at the EOC. We signed in as a resource with Gerald Pangelinan, the Resource Unit Leader. We learned that the radio room was being used by the Air National Guard. The HF antenna has been repaired, but is in use with the Guard. There is no place to put any antenna, and no place to operate a HAM radio at the EOC. The Radio Systems Administration team called the State EOC and found out that our task had nothing to do with the Bay County EOC, but only the State EOC in Tallahassee. We attempted to find actual need by briefly talking with the IC. but he was swamped with many inputs from a host of actors reporting on scene from government, law enforcement, fire and rescue, communications, industry, and private organizations. We repeatedly heard that there was no known request from the state EOC for amateur radio support at the EOC; only the Red Cross had requested support. Moreover, all the potential amateur radio facilities at the EOC had been re purposed for other more urgent needs. Consequently, no apparent request, need, or infrastructure for amateur support at the EOC. Met Gary Huckabay, recently arrived Red Cross supervisor, night shift. After discussion, it became clear that any remaining communication problems with the shelters could be solved by buying some AT&T "burner" phones. After further discussion with Gerald and Gary, it was determined that we would wait for a day to determine if our services were needed elsewhere (staging area for the County, for example), but with respect to the Red Cross, the need for our services was resolved: the Red Cross understood that there was a simple method for communicating with the shelters, and the EOC had no apparent need for or desire for amateur support. We made it quite clear that we capable of being re tasked to support other locations as needed, but this was a time consuming effort for us given the difficulties of road travel. Rutherford Red Cross shelter is being closed tomorrow (problems there), and Bozeman shelter can communicate by AT&T. As we had learned that the Red Cross current was moving admin operations to Bozeman and that special needs shelterees were being evacuated, having coordination in person with the Red Cross liaison at the EOC was crucial for our understanding our mission requirement and metrics for success. We asked Gary Huckabay about what to do about the requests for information about missing people in Bay County. We were told that they could look on line in the Red Cross "safe & well" site. We mentioned that these people did not have any access to the internet. It was then suggested that the relatives call the Bay County sheriff. In spite of the perception of a lack of need for communications, we noticed some disconnect between the information that the EOC was receiving and reality; for example, they were not aware that anyone had died at the Bozeman shelter. They also thought that the counts they received were exact numbers rather than estimates.

2200	Departed and returned to Bozeman operation.	
2300	Arrived Bozeman shelter. Very tired. Sent email to KX4Z apprising of the events and recording days events on 214.	
2400	Lights out.	

9. Prepared by (Name and Position)

October 15

UNIT LOG	1. Incident Name Hurricane Michael Panama City	2. Date Prepared 16 Oct 2018	3. Time Prepared 11:00 EDT
4. Unit Name/Designators Bozeman school shelter radio	5. Unit Leader (Name and Position) Shannon Boal		6. Operational Period 15 Oct 2018

7. Roster of Assigned Personnel

Name	ICS Position	Home Base
Leland Gallup	radio operator	Alachua Co
Shannon Boal	radio operator	Alachua Co
Susan Halbert	radio operator	Alachua Co

8. Activity Log

Time	Major Events

0700	Spoke with Karl Martin about the events of the previous day. There was a disconnect between what we had observed in Panama City and what people outside of the area perceived. Mr. Martin was under the impression that there was an operator at the Bay County EOC, and that the "Director of the EOC" wanted amateur radio communications. We attempted to explain that ATT phones were providing increasingly adequate communication, and that there was no amateur radio operator at the Bay County EOC to receive any shelter communications via amateur radio (nor was any such position likely to be staffed, because both the office and the equipment were in use by the National Guard!). Moreover, the IC did not need or want HAM radio communications. In reality, our assigned mission was no longer needed due to re-establishment of cell coverage; however, the conversation with Mr. Martin made an unpleasant end to an otherwise successful adventure.
08:00	Took down antennas and packed. This took about 4 hours.
12:00	Met with local shelter Red Cross shelter managers and received warm thank-you for showing up (never did meet the chiefs from National Headquarters).
12:15	Shannon fell on his back attempting to move a poorly secured pallet of waste cardboard placed in the road by some of the National Red Cross workers. He was injured, but not severely. Mr. Jeff, the chief custodian from the school, provided a bandage, disinfectant, water, aspirin, and Gatorade. There were staff changes occurring [which have been deleted in the interests of privacy. GLG]
12:30	Filled Susan's and Leland's cars with gas from our gas cans; left for Gainesville.
15:00	Stopped for lunch in Tallahassee (first meal of the day) at Steak & Shake. Limited menu, but AC and lights on. Gas is still scarce. Filled Susan's car with gas from our gas cans.
16:00	Found a station with diesel fuel for Shannon. They had no regular gas.
17:30	Arrived to the outskirts of Gainesville.

9. Prepared by (Name and Position)

5   KX4Z ICS-214

Note: Table formatting would not allow the ICS-214 to begin on this page.

ACTIVITY LOG (ICS 214)

1. Incident Name: Hurricane Michael	2. Operational Period: Date From: Time From:	Date To: Time To:
3. Name: Gordon	4. ICS Position: None (Alachua County volunteer)	5. Home Agency (and Unit):

6. Resources Assigned:

Name	ICS Position	Home Agency (and Unit)

7. Activity Log:

Date/Time	Notable Activities
Tuesday Oct 16 2018	**This report was compiled on Tuesday, October 16, 2018, in an attempt to record the recent events related to the Alachua County volunteers involved in Hurricane Michael.**
Saturday Oct 13 5 AM	I met 3 Alachua County volunteers at Exit 404 at 5 AM in the morning, to transfer fuel, chains, locks and other items to assist them as they headed out to the Panhandle. I had already loaned a generator and a Yaesu UHF mobile transceiver to Leland.
Saturday 7:30 AM	Days earlier I had sent press released out to local media related to our planned full scale exercise (S.E.T.) that would have been held on Sat. Oct 13; then I had sent press released notifying media it had been canceled. I wrote a new release that now we had sent 3 volunteers toward the Panhandle.
Saturday Afternoon	Cindy Swirko of the Gainesville Sun wanted their full names and I provided them but indicated Susan might be concerned about having her name printed and effectively that her home was vacant. In the end, the Gnv. Sun printed a story but just left all their names out, posted at 5:13 that afternoon. See: https://www.gainesville.com/news/20181013/local-ham-radio-operators-helping-with-post-hurricane-communications At 5:50 Kevin Bess emailed indicating it was a good release but might have been useful to run it through someone at the Section level. I replied that was a good idea but we would need to make it happen in timely fashion.
Saturday 7:13 PM	I had received an email from a member of my Sunday School Class about the Baptists needing radio support. At 7:13 PM I sent an email out to selected Alachua County friends and copied Karl Martin about Mr. Corbin's needs, some new email received by me mid afternoon. Copy of my email is appended. Nancy and I debated whether we could make it there to serve the Baptists but concluded we could not.

1. Incident Name: **Hurricane Michael**	2. Operational Period: Date From: Date To: Time From: Time To:
Saturday 7:29 PM	Jeff Capehart & I had been carrying on significant discussion about the mission requirements. In an email at 7:29 to Jeff, I quoted from Karl that the participants needed to have been background checked by their EOC; Jeff noted the discrepancy between that and an email that Jeff had received. My email is appended. Later I would see an email from Steve Szabo that the State would do the vetting. (so we were confused)
Saturday 8:59 PM	I had received an email from Karl Martin asking for clarification on the information I had sent to people in Alachua County about the need for help by the Baptists at Hiland Park Baptist Church. I replied at 8:59 PM, explaining we had no more people that met SEOC requirements so I felt obliged to see if anyone could help out the Baptists in need. My email is appended.
Saturday 9:15 PM	I had received a telephone call from the team in Bay County, and I wrote an email to selected Alachua County friends with the update and copied Karl Martin on it. My email is appended. I remember that I specifically left out some macabre details.
Saturday 10:03 PM	Received an email to me from Karl Martin indicating that the last contact he had had with the team they had reached a shelter and were planning to get Shannon to the EOC as an EOC station and then Susan and Leland planned to go back to shelter – Karl asked me to send him an email message to explain how to reach the team by WINLINK.
Saturday 10:52 PM	I replied to Karl Martin including precise instructions on how to reach the team using WINLINK. That email is appended, and includes a copy of Karl's request for instructions about their correct addresses.
SUNDAY 9:02 AM	Karl Martin replied to my email of 10;52 PM thanking me for the information.
SUNDAY AFTERNOON	Nancy and I visited our grandchild and son in Orlando on Sunday and on the trip back tried to listen to UHF repeaters of the SARNET. At approximately 3:30-4PM we heard Leland AA3YB many many times calling on the SAR-net trying to establish communications. The net control heard him many times and responded many times – it appeared that Leland could not receive the replies that were being sent to him. Nancy and I were concerned that there was a communications issue, distance problem, antenna problem, something.
SUNDAY 4:55 PM	I forwarded a message to Karl Martin that I had just found in Winlink email from the Alachua County team (they had sent it out at 20:36 GMT) The message is appended. In that email, Leland explained they had reached the Bay County EOC but found no radio operator there; that they were to return later in hopes of reaching the radio operator, and that they would keep Karl informed.
SUNDAY 5:07 PM	Leland send me an email over gmail explaining they were going to try to reach Hiland Baptist at the request of the Red Cross. That email is appended.

1. Incident Name: Hurricane Michael	2. Operational Period: Date From: Date To: Time From: Time To:	
SUNDAY EVENING	After I returned home from Orlando, i tried to monitor both SARNET and 3.950 on two radios in my living room. I was trying to listen to both SARNET and HF 3.950 simultaneously. At some point I believe i heard Susan's voice on one or the other. I gave up listening to the SARNET and tried to check in unsuccessfully --- so went upstairs to a better antenna/transmitter and checked in. There were discussions of why there were so few volunteers, the opinions expressed were it had to do with ICS requirements. I notified the Net Control Station N0TW that anyone could reach the Alachua County team using WINLINK. N0TW indicated that was not useful as he didn't have WINLINK. I was caught off guard and didn't know how to reply. Several minutes later I composed an explanation of how anyone using any email can email anyone using WINLINK by addressing CALLSIGN@WINLINK.ORG and putting //WL2K as the first item in the Subject line – and N0TW was very grateful for the explanation and quickly sent me an email to my WINLINK account to test this. I received his email and replied, giving all the Alachua County Winlink email addresses. There were others on the net who also seemed surprised that this could be done. As the misconceptions of the difficulties with volunteers continued on the net, I was able to explain that Karl Martin had put in writing all the exact requirements and that they were the First link on the section web page – but then i was told that people didn't know where the section web page was....I was again caught off guard and suggested they Google North Florida ARRL Section – at that point Kevin Bess came on frequency and gave the entire URL. At 8:13 PM i sent an email to Karl Martin documenting those events. The email is appended.	
SUNDAY EVENING 8:20 PM	I received a phone call from Karl Martin asking about what I knew about the Alachua County team's efforts and apparently in particular why they had visited the Hiland Baptist Church. We spoke for 20 minutes. I can't remember exactly what I told Karl but I looked it up and later that evening I sent him an email that I had received from Leland explaining they had been requested to do that by the American Red Cross	
SUNDAY 8:25 PM	I forwarded an email to Karl Martin from Leland (which I had received earlier) explaining why they had gone to Hiland Baptist Church. The email to Karl is appended.	

1. Incident Name: Hurricane Michael	2. Operational Period:	Date From: Time From:	Date To: Time To:
Monday 2:37 AM	I awakened in the night and found an email from Leland and forwarded it to Karl Martin. The email is appended. They described their efforts of the day a -- first visit to Bay County EOC – no radio operator, told he would be back at 7 Pm -- second visit to Bay County EOC – 8 pm, still no radio operator, but they figured out that they could communicate using AT&T phones. The email lists their points of contact. and finding out that all the requested posts in their area could communicate by simply getting AT&T telephones. My email to Karl is appended, which includes the email from Leland, written at 0053 AM Monday morning.			
Monday Morning at Work 09:08 AM	I got a phone call from Leland that they had been sent home.			
8. Prepared by: Name: _____ Position/Title: _____ Signature: _____			
ICS 214, Page 1	Date/Time: _____			

Saturday Oct 13 7:13 PM

Hello --

Sending this out to Alachua County- related emergency-communications hams.   Also to Marvin Corbin, from when it came, Already discussed with Karl Martin (Section Emergency Coordinator) who is still constrained by State rules [he took our word for our 3 volunteers and bent a regulation or two to get us in...] --- he sees nothing wrong with the following request from the Baptists (3rd largest disaster relief organization in the USA) . JEFF -- can you forward this to the entire yahoo mailing list and GARS mailing list if you feel appropriate?

I received (third or fourth hand) the request below from Florida Baptist Disaster Relief an hour ago.   I got back a phone call from Mr Marvin Corbin himself shortly after I tried to reach him.

SITUATION: Mr. Corbin is not an amateur radio operator and is trying to arrange for communications between a feeding station at a Highlands Baptist Church in Panama City, FL. He doesn't know exactly what he wants, but he wants the ability to have communications to the State EOC.   Since we have 3 volunteers already in Bay County who have WINLINK capability, and access to the FL EM NET on 3950/7251....those are pretty easy requirements to meet.

Mr. Corbin did not seem to have any other difficult requirements. *I believe he would be interested in anyone who could assist him.*

> ROAD CONDITIONS: https://fl511.com/#:Alerts   Mr. Corbin is very familiar with the roads. As you can see from that map on fl511.com, The PROBLEM is US. 231, the very last road to get to Panama City --- it is "one lane" due to downed poles.   Jeff Capehart says that there are crews aggressively clearing roads, so probably every few hours this gets better.   Mr. Corbin is emphatic that they are very much approved and you can very much help them out!!!

Had I known of this early this morning I might have been able to leave right then --- my hospital is very short (due to a national meeting) and I have to work Monday and have limited availability this week – but I have TONS of equipment and can assist anyone who wishes to go.   Mr. Corbin's email is above; you can contact him that way or through his telephone number: (352) 572 1227

IF YOU ARE INTERESTED – please contact Mr. Corbin **and also Me or Jeff Capehart** --- then maybe we can get a schedule going -- and maybe we could manage to staff these folks with the available manpower that we have. I may leave for there very early tomorrow morning with radios.   If I do, I'll leave an antenna up there.   if you are interested in going early tomorrow morning --- let me know.   I have to be back in Gainesville for work Monday.

Gordon Gibby KX4Z

-------- Original Message --------
Subject: FLDR Ham Radio Operators
From: Marvin Corbin <mailer@tpsdb.com>
Sent: Saturday, October 13, 2018, 4:40 PM
To: Shawna Puckett <shadowchik@hotmail.com>
CC:
TO: JEFF CAPEHART Saturday 7:29 PM

Karl quoted his own original request, in his NEW request for today, which has a link on the front page of the ARRL-NFL web page.

http://arrl-nfl.org/wp-content/uploads/2018/10/Hurricane-Michael-2018-Operators-Needed.pdf

here is a portion of what he stated (quoted from his original request):
  To anyone willing to deploy to the disaster area. We have an official request for operators from the state emergency operations Center. To be deployed you must fit the following criteria. • The 8 field operators will need to have this equipment. • **Must have been vetted (background checked) by their local county Emergency Management.** • Radio equipment for VHF/UHF (SARNET) HF (80M 40M 20M) and if possible HF Winlink • Antennas for radios • Alternate power (Solar, Wind, generator, etc.) • Food and Water for 7 days • Shelter • Supplies to sustain themselves in case there are no other resources available • Must be capable of deployment without any   [emphasis added by me]

Saturday 8:59 PM   My Email to Karl Martin:

Hi thanks Karl, we don't have any other people who can meet your requirements for service to the state, so I am making other Service opportunities known to people. Which is what I think ethics demands.

It turns out that the Southern Baptists are very near to where our volunteers were supposed to be, which means they could communicate through our volunteers.

However, our volunteers are not actually in Panama city because it did not work out well there, and they are all 3 at Bozeman which is north west of the city.   So they may not be able to assist the Southern Baptist.

They seem to be very happy with the help you were giving them, and they are doing the best they can.

I will not spread widely the choices you had to make in a hard time, and we hope to have this solved next time!

But you should know that there was a general feeling that the state was making this incredibly difficult, and that in the future we would be wise to have networking with other groups who might be more amenable to the kinds of service that we are able to provide outside of governmental systems.   The goal is to help disaster victims. In anyway that doesn't get in the way and fits into acknowledged groups.   I suspect you agree.

If you have any question, don't hesitate to contact me! Always willing to learn.

Gordon.

Saturday, 9;15 PM My Email to Alachua County friends and copied to Karl Martin

Passing along information received from our team tonight by telephone.

Incredibly difficult merely traversing small panama city; sirens everywhere. They think every policeman in Florida must be there! All three have settled for the night after 9 hours or more of travel, at Bozeman school northwest of Panama City after being told the need at another assigned shelter, which they reached, was not as great.

They have not yet got any antennas up. Pine trees snapped off in the middle; no tops. (Kinda what Steve Waterman always told me)   they are eyeing a tower they think they can exploit.

Bozeman shelter being used for special needs (extremely ill).   Of note, Allan West is deployed somewhere out there also with the FEMORS group.   Not sure where (forgot).

AT&T works right where they are, but extremely spotty. In town – much less connectivity.   Biggest problem they are addressing is that the county EOC seems to have no connectivity with Bozeman shelter so they are planning to tackle that tomorrow.   I read elsewhere that Verizon is working very very hard & steadily.   The team thinks that once phones work there will be much less need for them.

They were lucky to have cell phones with service. Future note; take cell phones from multiple vendors so if you're lucky, one will work!!

They are unsure whether they can make it back to assigned shelters inside Panama City.   Traffic, road conditions, etc.   Doing the best that they can.   Incredibly difficult to even move around modest distances.

But all safe, with real electricity and cots and comfort for tonight.

I don't have a phone number for Debbie Boals but I've checked with Angela Gallup and she has gotten word from her husband.

I wouldn't go sharing any of this with media......
Based on their 9 -hour trek to even get there....there is nothing that I can do with the one day I have remaining available.

Karl Martin has been doing great things and the team has really appreciated working with him.   I'm recruiting him to speak at our February Conference.

Cheers,

Gordon Gibby

Sure, karl!

Winlink email addresses are simply the callsign@winlink.org
So yours is KG4HBN@WINLINK.ORG.
I'm pretty sure that upper/lower case makes no difference in email addresses.

When you are initiating an email using WINLINK EXPRESS itself, it is even easier; there you merely need to type their CALLSIGN and the system assumes you mean their Winlink address. (its OK to put the entire @winlink.org if you wish, also)   Outside the system (like when using Gmail for example) you must fully specify KG4BHN@WINLINK.ORG

So their Winlink email addresses would be:

K4GLM@WINLINK.ORG
AA3YB@WINLINK.ORG
KG4VWI@WINLINK.ORG

from outside the Winlink system.

There is one more wrinkle to be familiar with --- the "white list" (list of people who are allowed to send an email to a winlink.org email account) -- because Winlink often goes over radio, they wanted to dramatically reduce the chances that people would be getting SPAM email.   So they created the concept of an allowed list.

1. If the Winlink user sends email to   whatever@whatever.com, that that person is automatically allowed to send email back to the Winlink user.   So if you want to be sure someone can reach your Winlink email --- just send an email TO THEM using Winlink Express and they automatically get added to the white list for your Winlink email
2. You can also put people on it manually, though I've never done it
3. You can get AROUND the white list by using the secret of putting //WL2K at the very first of your email SUBJECT (like I did above) --- something that spammers won't know.   So, had I initiated this email inside the Winlink system, I would not have needed it (you would have gotten it automatically) but since I initiated this email inside of GMAIL, I needed to use that trick to make sure that your WINLINK account received this email.

Hope all that helps!!

gordon gibby

On Sat, Oct 13, 2018 at 10:03 PM Karl Martin <kg4hbn@gmail.com> wrote:
> Gordon, Could you send there win link email addresses to my Winlink? kg4hbn@winlonk (.net?. .com?) Last I spoke to them (cell phone) They had got to the shelter and was going to drive Shannon to the EOC to setup as the EOC station then Leland and Susan was going back to the shelter.

SATURDAY 10:52 PM TO KARL MARTIN

SUNDAY 4:55 PM
following message just received from Bozeman. They sent it to your WINLINK address as well, but to be certain you get it I'm forwarding it to your SMTP: address.

Message ID: W7DTLR4O76O0
Date: 2018/10/14 20:36
From: AA3YB
To: KG4HBN; bob.holdredge@redrcoss.org; cesar.rivera@redcross.org
Cc: W4UFL; KX4Z; gallupleland@gmail.com
Source: AA3YB
Downloaded-from: Telnet:cms.Winlink.org
Subject: //WL2K Amateur Radio Communications Established at Bozeman High School Shelter P

The Alachua County amateur radio ARES team has established HF, VHF, and UHF communications at the Bozeman school shelter, Panama City, FL..

Approximately 100 special needs persons at this shelter; operations by the medical personnel have been grim.

We are able to and have sent and received digital message traffic by HF Winlink. Have already received requests to pass health and welfare messages about persons in Mexico Beach. We are unable to do so because no communications with Mexico Beach, but will pass when able to the County EOC.

We have established communications with the North Florida ARES net by HF phone, and with the SARNET by UHF phone.

We have established communications locally through a now-operating repeater and have passed health and welfare traffic.

We visited the Bay County EOC to liaise with the amateur radio operator we were told was assigned there; we found that was not present but would return at 1900 local ; we told the EOC point of contact that we dealt with that we would return this evening. We also left a note with our simplex VHF frequency and AA3YB@winlink.org as a digital Winlink address for traffic. Local road traffic is very difficult, with massive debris and many emergency vehicles.

We have coordinated with the ARC shelter director, Mr. John Sanders, that we have email and phone capability and are able to pass messages to ARC, etc No traffic yet; we will repeat so that the Shelter is aware of our essential comms service, and that we can pass health and welfare traffic as priority allows.

Will keep you apprised of our operations.

Leland Gallup

SUNDAY 5:07 PM

FROM LELAND:

Gordon, we are responding to a request from the state American Red Cross through the sarnet to go physically to Highland Park to assess their Communications needs. We're off to do that now. Will advise on 3950 sometime after 7 as to results.

SUNDAY 8:13 pm

To Jeff & Karl Martin
When I couldn't take it any more – the difficulty the folks on 3950 were having trying to reach Leland, Susan, or Shannon --- I fired up the SHARES station on ham frequencies and tried to explain that they could easily reach them on WINLINK

To my horror, the net control indicated that they didn't know WINLINK and so that wouldn't help them....

So then I realized I have not explained it correctly....so very nervously I came back in again later and asked to explain to them how ANYONE WITH EMAIL could reach the Alachua County team using Winlink....they were very willing for me to explain that so I nervously explained how you address email to

AA3YB@WINLINK.ORG

and you put //WL2K as the first item on the subject line to get through the white list.

The net control station, N0TW is apparently a fast learner because he asked for MY Winlink email and he sent ME a Winlink email --- which he did correctly, so I got it, and replied to him again giving him the addresses of all the members. Terry, N0TW just emailed me that he has printed out the information on how to do it.

Needless to say....this is emergency training that people need....

Another comment: There seems to be huge misunderstanding on why so few volunteers were able to volunteer...I heard people citing all kinds of mistaken ideas.....the real problems in Alachua County were

1) none of us have been background checked "by the EOC"
2) very very few of us can leave our jobs for 7 days

I tried to explain #1 and people seemed to understand that, and the correct news that Phil Royce is working for Volunteer America (or someone else) to get the background checks done was brought out on the net by another person. Later in the net, I was able to politely correct and give people direction to go to the NFL Section Web page where the exact information is printed. The Section Manager also checked in and gave the exact web address, very helpful. This kind of information simply had not been getting out to the net control stations of that net.

There was also great confusion about what the requirements were for people to assist the Florida Baptist Disaster Relief Team --- people did not seem to understand that this group is separate from ARES and has different volunteer requirements (very few requirements). I'm hoping that we will build far better relationships with these NGO groups performing valuable service and also able to accept ham radio volunteers that might not meet the requirements of the State EOC and therefore aren't going to be otherwise used....

I hope that all of these issues that have occurred will be collated into a great After Action Report / Improvement Plan so they will get SOLVED.   Our group has published now three such reports, the first of which had 36 issues....most of which we got solved.

Thank you so much for your work,
Gordon Gibby

SUNDAY 8:25PM

TO KARL MARTIN

This is the email from Leland explaining that the American Red Cross asked them to do that.

---------- Forwarded message ---------
From: Leland Gallup <gallupleland@gmail.com>
Date: Sun, Oct 14, 2018 at 5:07 PM
Subject: Hiland Baptist Recon
To: Gordon Gibby <docvacuumtubes@gmail.com>, Gordon Gibby <ggibby@anest.ufl.edu>

Gordon, we are responding to a request from the state American Red Cross through the sarnet to go physically to Highland Park to assess their Communications needs. We're off to do that now. Will advise on 3950 sometime after 7 as to results.

MONDAY 2:37 AM
My Email to Karl Martin

Karl, I woke up at 2:30 and read this long email from Leland (below) sent just before 1 AM   Shannon, Susan, Leland are back at the Bozeman school. They have apparently managed to teach personnel at three locations how to have good communications by simply buying AT&T cell phones: Hiland church, Bozeman, and Bay county EOC.

Leland is a lawyer by trade and he likes words, so he uses lots of them in the message below ; I had to read it three times to put it all together.

After you figure it all out, suggest that you quickly send him / them directives of where to go & what to do next. — they appear to be rapidly solving communication problems for the facilities they have visited so far, despite arduous travel times.

Please let me know if I can assist you in any additional way.   I will be monitoring from the operating rooms at Shands in the morning, but I don't have radios there.

Gordon Gibby
352 246 6183
Docvacuumtubes@gmail.com

Sent from my iPhone

Begin forwarded message:
> **From:** Leland Gallup <gallupleland@gmail.com>
> **Date:** October 15, 2018 at 00:53:23 EDT
> **To:** Gordon Gibby <docvacuumtubes@gmail.com>
> **Cc:** Leland Gallup <gallupleland@gmail.com>
> **Subject: Re: news from Leland Susan and Shannon**

Gordon, the three of us have just returned from the Bay County Emergency Operations Center and returned to the Red Cross shelter at the Deane Bozeman School north of Panama City.

> Before I get into what happened today, one of the things I would like to underscore is that is important to understand the state of confusion/semi-controlled chaos in a county Emergency Management operation overwhelmed by calamity of this scale and trying to coordinate resources flowing in from all over the country – government and the private sector.
>
> Next point I want people to understand is that doing anything in this County by way of travel from point A to point B is also very very time-consuming right now. For example, it took us one and a half hours today to go approximately 7 miles from our location at the Deane Bozeman High School to the Hiland Park Baptist Church, and then two hours to drive ten miles (!) From the Hiland Park Baptist Church to the Bay County EOC building. Think about that.
>
> Doing anything that involves getting in an automobile and going anywhere is an ordeal. Trust me on this, I know. So did my two passengers. but, as we have learned from this exercise, it was absolutely critical for us to go physically to places to really and truly establish the need for communications, or not, and to cut through layers of bureaucratic confusion, misunderstanding, and miscommunication to again determine what actual need for communications existed that could be met most effectively by amateur radio operators: I stress – by amateur radio operators.

Anyway to make a very long story short about today. We went to the EOC earlier in the afternoon, but left when we found that "the ham" would be back at 1900. It was then that I sent you the text.

When we arrived back at the EOC at approximately 2000 we found that the ham wasn't a radio operator, but a county employee who just happened to be a ham....but not working as an operator.; in fact there was no operator, and our information was to this point confused and uninformed.

It was only after discussion with the director of the Red Cross at the Bay County EOC, and the director of resources for the Bay County Emergency Operations Center, that we determined the following:

There is no real need for our support to the Red Cross as per our original support request from them. Please do not put that out tonight, because that isn't official.

The fact is that all the Red Cross needs for effective communication between shelters and EOC- Red Cross liaison at the EOC can be met by Red Cross people buying AT&T burner phones from the nearest AT&T store...at least as to what we can see in Panama City.

That's because AT&T had cellular Mobile units set up at various locations that actually provide communications between, for example, the shelter or relocated in the OC. Verizon and Sprint, not so. That is changing, and I suspect soon, given the flow of disaster response descending on the county we are in.

So the answer to communications needs that we were expected to meet was so simple it was startling. Buy AT&T backbone phones and your problem is solved. When we said that to the Red Cross director the light came on in his eyes, and they frankly acknowledged that yes you're right we can do it with AT&T phones. Why do we know this? Well, because I can use my own AT&T phone and have a very strong signal at the EOC and at the Bozeman school where the shelter is located.

After we determined that they really didn't have a need for ham radio operations, the Red Cross said okay, well we really don't need you. Then we discussed with the Resources Director for the Bay County collective operations, who is coordinating the vast number and sources of resources just how amateur radio operators might support the County's enormous emergency operations. And those operations, frankly, are gigantic and overwhelming them. I have a do-out with the resources director to determine whether the County needs hams to provide link between a new staging area and the EOC. I expect that determination some time tonight. If the need exists that will be routed through the State EOC.

Our view is that cell service is rapidly coming on line; so actual communication by amateur radio --- our opinion by seeing ground truth and educating a number of people on what we do and how we do it-- is PROBABLY not really necessary in Panama City proper, at least as that support is provided to our conventionally served agencies and organizations. Other places not high on the cell providers priority list for restored service? Maybe more so.

I also concluded that supervisors at the EOC were overwhelmed by trying to span control over a bazillion number of requests from various people for direction and that it was hard to get directive authority from the very stressed directors. So Susan and Shannon were very put out by our reception by the IC; too many stressed people running around. The resources coordinator has been good to us, on the other hand, because he was lower in the food chain and could give more undivided attention.

I also need to touch on the Southern Baptist communication facility at Hiland Park. impressive! But they didn't have a working HF operation. We helped a ham there get on the local VHF repeater. They are a feeding operation distribution point facility, not a shelter.

I am too tired now, but if you can pass to Karl Martin that they feel that they still need ham support, that box is checked. But again, quite frankly, I think the cell providers will rapidly address that need. This is not Haiti.

If we learn from the EOC that we aren't really needed, we will close down and return to home station.

That's it for now. We are exhausted. Not much radio work, but a whole lot of time in vehicles and determining actual need. That's actually what these operations actually mean in the year 2018 in the United States, given the disaster industrial complex we have developed.

In my view the real need for amateurs is in the 48 hours after calamity; by the time we got our marching orders and set up, communication needs were rapidly being addressed...even if we had to advise people of that emerging fact.

We have learned much from this event. And much of that had to do with sorting out facts, dealing with lacking interpersonal communication in a complicated inter-agency operations environment, and just the attending to the physical demands of finding the right people to talk to at the right places, and that just took time in a vehicle.

That's it for now.

More tomorrow as we emerge from the fog of bureaucratic fug.

Leland

(This page intentionally blank.)

6 SUSAN HALBERT'S NARRATIVE

Alachua County Hurricane Michael HAM radio response
Susan Hilbert's (KG4VWI) narrative

Time line (plagiarized liberally from ICS214 forms)

05:00 rendezvous at Exit 404; obtained gasoline, other supplies from Dr. Gibby; made plans

08:00 stopped in Tallahassee for gas and food; Waffle House: good waffles, nice waitress; gas hard to find; Waffle House limited menu; Waffle House index yellow.

12:30 Arrived at Northside Elementary Shelter; ATT phones working intermittently;, no AC, no water or sewage. Shelter manager (Angie) suggested we go to Bozeman, where there was greater need (no communications). Took one vehicle to scope out Bozeman school.

13:30 Arrived at Bozeman shelter (K-12 school). 2 people from Lee Co EOC met us with animated request for communication support (Caitie Eck, John Kelly). They thought that we were the folks that they had ordered that morning. We explained that we were ordered via the Red Cross for shelter support. John Sanders 'Sandy' was concerned that these orders be deconflicted, because he was afraid that when the state saw that there were two similar orders, one would be canceled. We called Karl Martin, our ARES supervisor. He was made aware of the two similar orders, and that there was a need for both. He agreed that we should stay at Bozeman. We were also asked to check out the Bay Co EOC; Shannon had requested posting at the State EOC, so if the Bay Co. EOC needed help, he could go there.

14:00 We proceeded with arrangements for operating facilities at Bozeman school. Kim Timmons, the assistant principal, helped us. We were not able to use the communications room because it was being used as a temporary morgue; however, there was a room nearby with ideal facilities. It has running water, power, and even a shower. "Mr. Jeff" (father was W4TAH), the school handyman, found us cots, bedding, and other supplies. We had received several versions for the address of the Bay County EOC. Kim knew where it was and gave good directions. It is not far from here (7 miles).

15:00 Left Bozeman to collect our vehicles, reconnect with Angie at Northside. She agreed with us that we should go to Bozeman. Traffic and debris made travel difficult all day. Roads reduced to one lane each way or less; side roads were impassable or filled with vegetation, downed wires, and other hazards; parking lots were infested with nails to puncture your tires (no mishaps for us!).

18:30 Returned to Bozeman. Unloaded equipment. Too late (and dark) to put up antennas.

19:30 Checked in with night shelter staff.

20:00 Checked in with Gordon Gibby.

14 Oct 2018
7:30 Set up stations. Push-up pole on a tripod with UHF, VHF station at the top, US flag, Buckmaster antenna on a pulley.

10:08 Established UHF phone communications with SAR-net through the Chipley station at 10:08 CDT. We were their first Bay County contact.

10:30 Established HF phone communications with NFAN on 3950 kHz.

11:00 Established HF digital Winlink communications on 40 meters. Sent one message; received 7.

11:25 Replied to one of the health and welfare messages.
Received request to check on Hiland Baptist Church, from Red Cross, to assess their communications needs.

11:35 Received message from Gordon that WINMOR was not working, so test message sent on VARA.

11:35 Established VHF phone connection with local repeater (later learned on 800 ft. tower) 145.330, tone 100. We listened quite a bit on this repeater and learned that there was no contact with the Bay Co. EOC.

11:30 Touched base with the staff at Bozeman and; the Health Department staff (special needs shelter) said it was a horrible night. Staff we spoke with at shelter said that they had "no communications" with the EOC. This explained why we had been tasked to provide communications with the EOC from the shelters (shelter counts, needs, etc.). ATT phones were working, but possibly there often was not a response from the EOC, and routine messages had to be sent by courier, which could be a 2 hour round trip. It seemed to us that in order to accomplish our task, it was essential to find out how to obtain communications with the local Bay County EOC.

12:30 Went to EOC to figure out how to obtain communications there. This took considerable time on road; we made decision to have all three go with the possibility of leaving one of us there at the EOC to provide the communications link with Bozeman. (Karl Martin had suggested posting Shannon there.) We could not find a person at the EOC able to identify who we should talk to about radio communications at the EOC; the "ham" we were informed was there in fact was not there; we were told to return at 19:00 when Bob Edmunds, the EOC operator, would return for the night shift. LEO on duty said that their radios mostly didn't work. ATT cell phones were the best option.

12:30 Continuing to make contacts on our station. Shannon went to see the local Bozeman Red Cross staff to advise them that we were going to the Hiland Baptist Church. Mr. Sanders was not available, but Shannon spoke with James "Doc," a Red Cross worker. Doc went to a "no admittance" office while Shannon waited a long time but never obtained access. (He learned that the Red Cross was making the Deane Bozeman school their primary location, hence the private office.) The Red Cross was moving Special Needs patients to Tallahassee to make room for their headquarters at the school. People were arriving from Washington DC. We again were asked again to go to Hiland Baptist Church to assess communication needs.

14:30 Went to Hiland Baptist Church; arrived about 16:00. This took one and a half hours on the road. The Baptist Church has a dedicated high end communications trailer, equipped with a HAM radio that will do all bands (ICOM 706 Mark II G). Tim (KD5SSF) is an experienced operator with the Southern Baptist Disaster Relief, especially on HF, but there was no HF antenna. We helped him set up the radio to communicate with the local 145.330 repeater. We also told them they could probably reach us directly. The Hiland operation was a food distribution operation, not a shelter. Having satisfied the requirement we were given, we had supper and left.

18:15 Left Baptist Church to go to the EOC. It took 2 hours to drive there due to bad traffic. Roads were hazardous due to downed power lines and poles, pieces of demolished buildings, broken glass, large amounts of vegetation from broken trees, convoys of emergency vehicles, traffic accidents, non-functional traffic signals, and general traffic congestion.

20:15 Arrived at EOC.
Met Charlie Wooten (NF4A) by coincidence, as he was leaving.. He owns the 145.330 repeater and runs several local radio stations. He has been in the local HAM community for many years. He is somewhat discouraged by the lack of interest in emergency communications in the local HAM community. He was not at the EOC as an amateur operator, but instead in his professional capacity with responsibility to have local radio stations up and running. (Learned tonight on NPR – 22 October – that the local PBS station has set itself up in the EOC to provide public information.)

Met Bob Edmunds, the HAM who had checked into the 3950 net from Bay County. He works for the EOC, but he is assigned to GIS and traffic, not radio. There was no known amateur radio operator or any ability to operate at the Bay County EOC at that time.

We signed in as a resource with Gerald Pangelinan, the Resource Unit Leader. We learned that the radio room was being used by the Air National Guard. The HF antenna has been repaired by the National Guard, but is in use for their mission. There is no place to put any antenna, and no place to operate a HAM radio at the EOC. The Radio Systems Administration team called the State EOC and found out that our task had nothing to do with the Bay County EOC, but only the State EOC in Tallahassee. We attempted to find actual need by briefly talking with the IC. but he was swamped with many inputs from a host of actors reporting on scene from government, law enforcement, fire and rescue, communications, industry, and private organizations. We repeatedly heard that there was no known request from the state EOC for amateur radio support at the Bay County EOC; only the Red Cross had requested support. Moreover, all the potential amateur radio facilities at the EOC had been re purposed for other more urgent needs (National Guard). Consequently, no apparent request, need, or infrastructure for amateur support existed at the EOC.

Met Gary Huckabay, recently arrived Red Cross supervisor for the night shift. After discussion, it became clear that any remaining communication problems with the shelters could be solved by buying some AT&T "burner" phones. After further discussion with Gerald and Gary, it was determined that we would wait for a day to determine if our services were needed elsewhere (staging area for the County, for example), but with respect to the Red Cross, the need for our services was resolved: the Red Cross understood that there was a simple method for communicating with the shelters (AT&T phones), and the EOC had no apparent need for or desire for amateur radio support. We made it quite clear that we were capable of being re tasked to support other locations as needed, but this was a time consuming effort for us given the difficulties of road travel. Rutherford Red Cross shelter is being closed tomorrow (problems there), and Bozeman shelter and the Northside shelter can communicate by AT&T.

We asked Gary Huckabay about what to do about the requests for information about missing people in Bay

County. We were told that they could look on line in the Red Cross "safe & well" site. We mentioned that these people did not have any access to the internet. It was then suggested that the relatives call the Bay County sheriff.

In spite of the perception of a lack of need for communications between the shelters and the EOC, we did notice some discrepancies between the information that the EOC was receiving and reality; for example, they were not aware that anyone had died at the Bozeman shelter. They also thought that the counts they received were exact numbers rather than estimates. These problems should resolve as the AT&T network becomes more robust.

22:00 Departed EOC and returned to Bozeman operation.

22:30 Arrived Bozeman shelter very tired. Leland sent email to KX4Z apprising of the events and recording days events on 214.

24:00 Lights out.

15 Oct 2018
07:00 Spoke with Karl Martin about the events of the previous day. There was a disconnect between what we had observed in Panama City and what people outside of the area perceived.

Mr. Martin was under the impression that there was an operator at the Bay County EOC, and that the "Director of the EOC" wanted amateur radio communications. We attempted to explain that AT&T phones were providing increasingly adequate communication, and that there was no amateur radio operator at the Bay County EOC to receive any shelter communications via amateur radio (nor was any such position likely to be staffed, because both the office and the equipment were in use by the National Guard!). Moreover, the IC did not need or want HAM radio communications. In reality, our assigned mission was no longer needed due to re-establishment of cell coverage; however, the conversation with Mr. Martin made an unpleasant end to an otherwise successful adventure.

08:00 Took down antennas and packed. This took about 4 hours.

12:00 Met with local shelter Red Cross shelter managers and received warm thank-you for showing up (never did meet the chiefs from National Headquarters).

12:15 Shannon fell on his back attempting to move a poorly secured pallet of waste cardboard placed in the road by some of the National Red Cross workers. He was injured, but not severely. Mr. Jeff, the chief custodian from the school, provided a bandage, disinfectant, water, aspirin, and Gatorade. [GLG: there were other staff changes noted here but in the interests of privacy they have been deleted.]

12:30 Filled Susan's and Leland's cars with gas from our gas cans; left for Gainesville.

15:00 Stopped for lunch in Tallahassee (first meal of the day) at Steak & Shake. Limited menu, but AC and lights on. Gas is still scarce. Filled Susan's car with gas from our gas cans.

16:00 Found a station with diesel fuel for Shannon. They had no regular gas.

17:30 Arrived to the outskirts of Gainesville.

Some thoughts from Susan
This deployment was a tremendous learning experience. I have never gone on a multi-day radio deployment

outside of Gainesville before. Packing and planning quickly, the confusion about whether we were to go or not, the confusion about the mission – all of these difficulties are part and parcel of any emergency. It was good experience.

We were able to get a high quality station up and running within a few hours. It took all of us. It would have been very difficult to do alone. We were able to contact SAR net (closest station in Chipley), the local repeater, and the 3950 net. We also were able to send and receive WINLINK email over HF. Each person's skills were essential to success.

Did we help anyone? We did in fact help the Baptists to get their station running. They will be using their radio for communication between headquarters and strike teams providing real help in areas not reached by government agencies. That is a valid use for HAM radio, even when cell phones in urban areas are working increasingly well.

Probably we arrived too late to provide much help for communication between the shelters and the EOC. ATT cell phones were coming on line very quickly. Moreover, as we later learned, there was nobody at the local EOC to respond to whatever we might have requested. Traditional shelter communication help is needed during and just after the storm, before cell phones are working. It is contingent on a strong amateur radio presence at the EOC, or at least access to local emergency management personnel. If we had put up the antennas on Sunday night, there is a small possibility that we might have been able to relay some urgent messages from the special needs shelter, but there would not have been anyone local to respond.

We had determined on Sunday night that the mission assigned to us – to support Red Cross shelter communication – was no longer needed. The Red Cross liaison in the EOC realized that he did not need us because he could buy ATT cell phones that worked. There was little possibility, given the bureaucracy, and the lack of first-hand knowledge on the part of our supervisors, of getting us assigned to another task. Thus, the decision to send us home was the correct one, albeit for the wrong reasons.

Another complication that we heard about after the fact was that the US President was due to visit Panama City in a day or so. Bozeman school is located outside of town to the north, on the way into town from I-10. If the President were to visit a shelter, that would be the one to visit, avoiding many of the hazards and traffic problems present in the city itself.

There is a common saying in emergency management that all disasters are local. When the need is greatest, outsiders cannot get there. Local resources have to be in place. For traditional shelter communication support to happen effectively, local radio operators must be in place at the shelters and at the EOC ahead of the storm. In a few days, when they get tired, by then people from elsewhere can get there to help local people if there is still a need. In the case of Panama City, there were not local operators in place.

In my experience, much of emergency management is being prepared (planning) for things that do not develop as planned. Several times in the past, I have had all my gear ready to go to a shelter that never opened. This is the nature of emergencies. They are unpredictable, and in fact, that's what makes them emergencies. In this case, we gathered all the items and drove all the way to Panama City, only to turn around and come home when we found that the mission we were assigned to was not needed. This is the nature of emergencies. We practice for a lot of things that (thank God) have not happened.

Tonight (22 October, one week after we came back to Gainesville), there was a story on NPR about how there was still no communication infrastructure in Panama City. Verizon was mentioned by name, indicating that probably their cell phones still do not work. People did not have cell phone availability, so they could not call their insurance companies, tree services, employers, etc. Many people were still listed as missing because they had no way to contact anyone. Had our handlers, supervisors, and managers been more flexible there is a possibility that we could have been reassigned to a useful task, such as providing communications for search and rescue task forces, like the Baptists planned to do. However, given the stresses involved in a major disaster, thinking creatively about deploying volunteers you never saw before in your life would not come easily. Nobody had ordered more HAM radio operators, and none of the local managers perceived a need. I do not regret having gone, and I do not fault anyone for the outcome. It was a good experience overall. I learned a lot.

A few lessons (no order):
1. It is necessary to have a team.
2. Flexibility is crucial; miscommunications, misunderstandings, emotions are part of any emergency.
3. Remember always to look at the big picture. Does HAM radio add anything to the solution, or is there a better way to solve the problem?
4. Always be sure to keep supervisors apprised of the local situation (we did poorly at this).
5. Bring an antenna that does not require support from trees or structures.
6. Take plenty of flagging tape; invest also in some caution tape.
7. Take a variety of means of anchoring the antenna installation.
8. A printer (battery operated) could come in handy.
9. **Two is one; one is none.**

ABOUT THE NORTH FLORIDA AMATEUR RADIO CLUB

The North Florida Amateur Radio Club was formed in order to better support the ARES(R) mission in Alachua County. The formation of the club allowed acquisition of a club callsign and also liability insurance – two crucial assets when carrying out simulated exercises on public or private property with permission, and when using WINLINK radio email (which requires callsigns for its own email addresses).

The club maintains a web site (https://www.qsl.net/nf4rc/) and is very active in carrying out NIMS-compliant exercises and writing them up afterwards, in HSEEP format when possible. These are published on the club website and usually also on Amazon as soft-cover books.

Previous publications of this club include:

2017 Hurricane Exercise
https://qsl.net/nf4rc/2017AlachuaCountyCreateSpaceAfterActionReport.pdf
https://www.amazon.com/Alachua-County-Hurricane-Action-Reports/dp/1548062200

2017 Steinhatchee Storm Exercise
https://qsl.net/nf4rc/2017AlachuaCountyCreateSpaceSteinhatcheeAAR.pdf
https://www.amazon.com/Steinhatchee-Storm-How-Puerto-Rico-volunteer/dp/1978441509

2018 Wacassassa Wildfire
https://www.qsl.net/nf4rc/2018/2018 AlachuaCounty Wacassassa Wildfire Excersize.pdf
https://www.amazon.com/Waccasassa-Wildfire-Exercise-Alachua-Reports/dp/1721727817

www.ingramcontent.com/pod-product-compliance
Lightning Source LLC
Chambersburg PA
CBHW062337220526
45469CB00008B/2750